W9-BFI-182

Vegetables

Ann Thomas

Philadelphia

This edition first published in 2003 in the United States of America by Chelsea Clubhouse, a division of Chelsea House Publishers and a subsidiary of Haights Cross Communications.

Chelsea Clubhouse
1974 Sproul Road, Suite 400
Broomall, PA 19008-0914

The Chelsea House world wide web address is www.chelseahouse.com

Library of Congress Cataloging-in-Publication Data

Thomas, Ann, 1953-
Vegetables / by Ann Thomas.
p. cm. — (Food)

Includes index.
Summary: Presents information on the importance of the vegetables food group, describing various kinds of vegetables and how they are grown, handled, and prepared.

ISBN 0-7910-6977-X
1. Vegetables—Juvenile literature. [1. Vegetables. 2. Nutrition.] I. Title. II. Food (Philadelphia, Pa.)
TX557 .T46 2003
641.3'5—dc21

2002000033

First published in 1998 by
MACMILLAN EDUCATION AUSTRALIA PTY LTD
627 Chapel Street, South Yarra, Australia, 3141

Text design by Polar Design
Cover design by Linda Forss
Illustrations © Anthony Pike

Printed in China

Acknowledgements
Cover: Getty Images

Kathie Atkinson/AUSCAPE, p. 5; Australian Picture Library, pp. 12 ©Gerry Whitmont, 19; Coo-ee Picture Library, p. 17; Great Southern Stock, pp. 13, 20, 16, 26, 28, 29; HORIZON International, pp. 8, 9, 11, 18, 22, 25; Photolibrary.com, pp. 6 ©Jenny Mills, 14, 22 ©Benelux Press, 23 ©Donna Day, 24 ©Chris Everard; Stock Photos, pp. 15 ©Lance Nelson, 20 ©Rick Altman, 27 ©Dick Luria; U.S. Department of Agriculture (USDA), p. 7.

Contents

Why Do We Need Food?

We need food to keep us healthy. All living things need food and water to survive.

Echidnas feed on termites. They live in Australia and New Guinea.

There are many kinds of food to eat. People, animals, and plants need different types of food.

What Do We Need to Eat?

Foods can be put into groups. Some groups give us **vitamins** or **minerals**. Some groups give us **proteins** or **carbohydrates**. We need these **nutrients** to keep us healthy.

We need to eat a variety of foods.

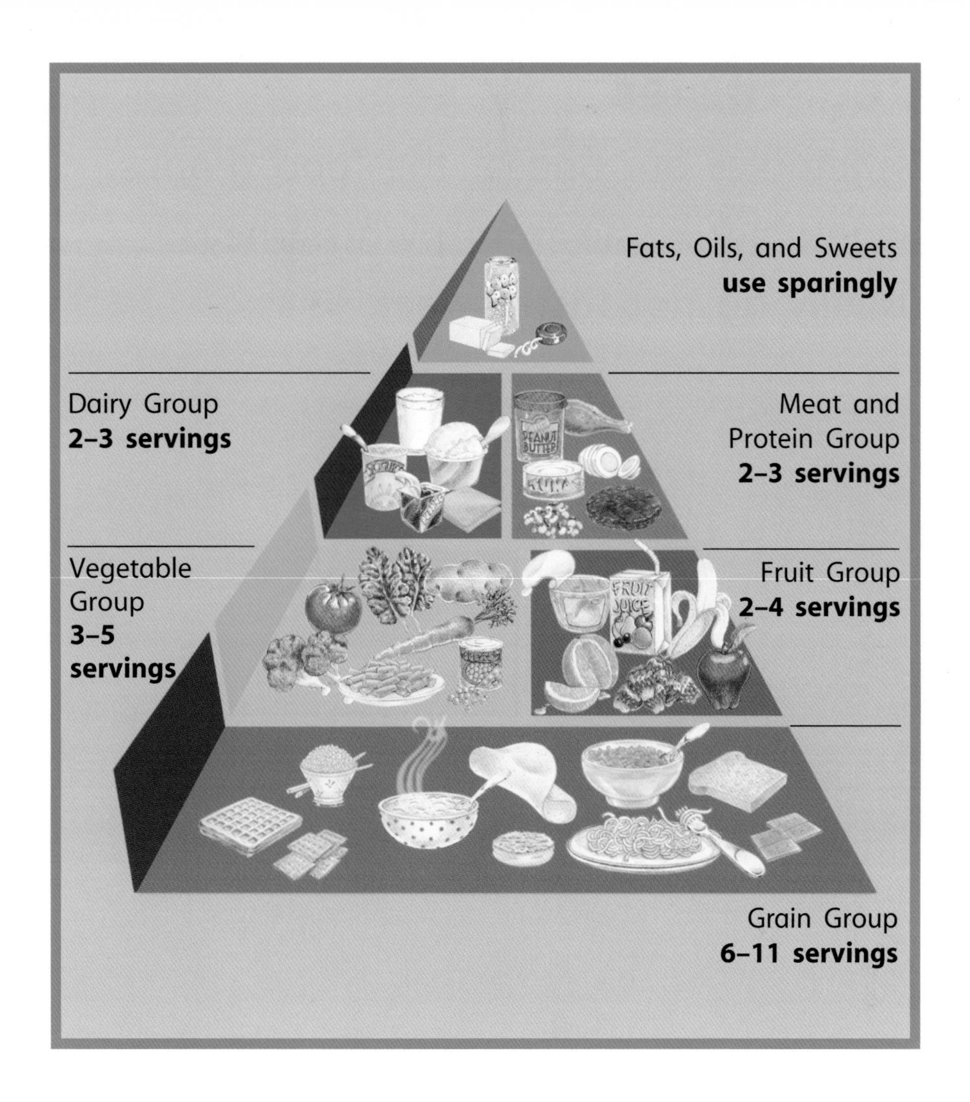

The food guide pyramid shows us the food groups. We should eat the least from groups at the top. We should eat the most from groups at the bottom.

Vegetables

Vegetables are one of the food groups. Vegetables come from plants. They usually do not taste sweet.

Vegetables contain vitamins, minerals, and carbohydrates. These nutrients help us grow.

What Are Vegetables?

Vegetables can be different colors. There are green, yellow, orange, red, purple, and white vegetables.

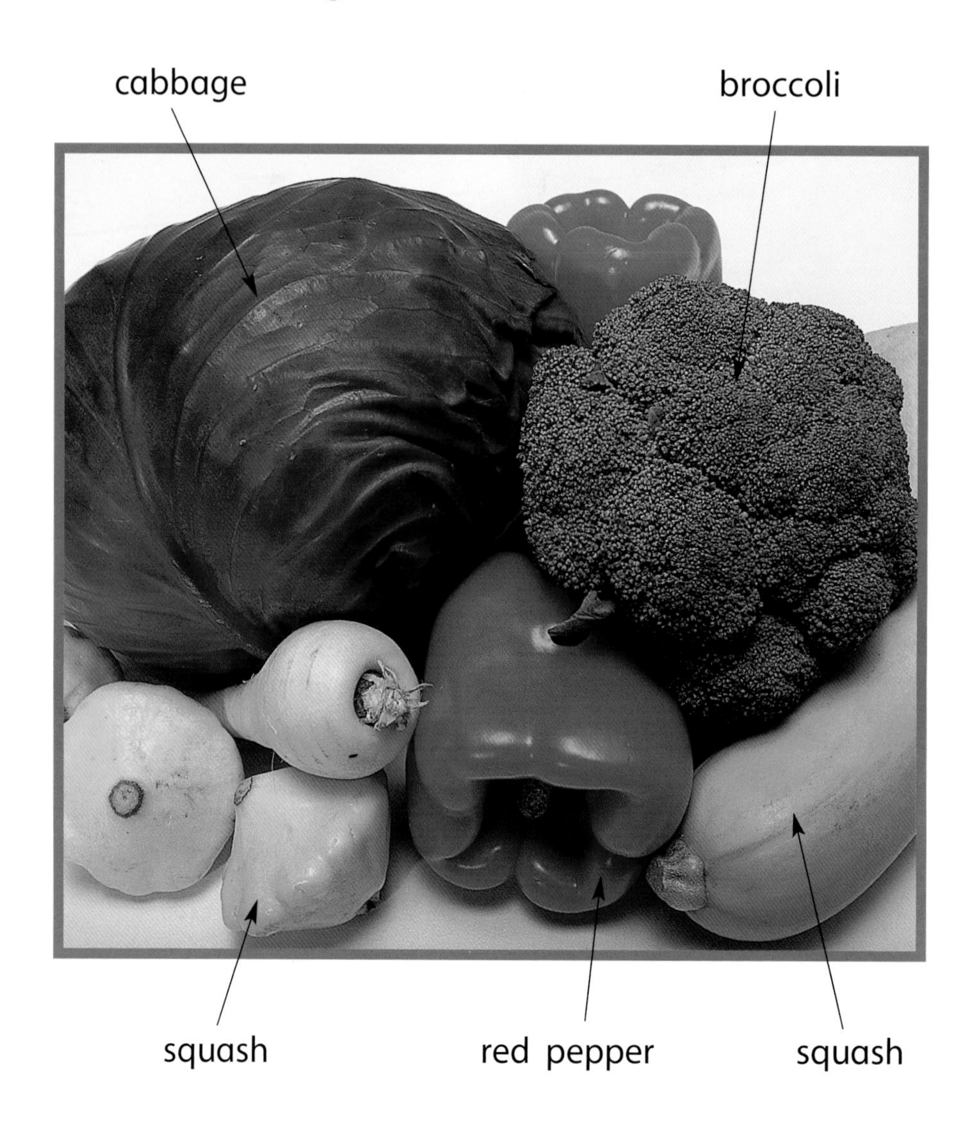

Green beans grow above the ground.

Vegetables grow in many places and on many parts of plants. Beans grow on plants above the ground. Beets grow underground.

Spinach and lettuce are leaves of plants. Carrots and radishes are roots. Potatoes are tubers. Tubers are a kind of stem that grows underground.

Spinach contains many vitamins and minerals.

Peas grow inside shells called pods. Some pea pods are tender enough to eat.

Peas and corn are seeds. Celery and asparagus are stems. Onions and garlic are **bulbs**.

Growing Vegetables

Some vegetables grow best when the weather is warm. People plant sweet corn in late spring. They pick sweet corn in late summer and fall.

People eat the white flower buds on the cauliflower plant.

Some vegetables grow best when the weather is cooler. Onions, cauliflower, and cabbage are planted in early spring and picked in early summer.

Most vegetables grow from seeds. Some seeds are large and some are small. Some are soft and others are hard.

People grow vegetables on farms, in gardens, in pots, and in **greenhouses**. Greenhouses protect plants from bad weather.

People pick vegetables carefully to keep them from bruising. On large farms, workers pack vegetables in boxes or bags. Trucks carry vegetables from farms and greenhouses to supermarkets and restaurants.

Some farmers grow many kinds of vegetables in large gardens. They sell their vegetables at markets or to restaurants or stores. The vegetables are organic if the farmers do not use **chemicals** on the plants.

Buying Vegetables

Many people buy vegetables at large supermarkets. Supermarkets have a wide variety of vegetables. Some vegetables come from local farmers. Other vegetables come from places far away.

Supermarkets stock large amounts of vegetables.

Some small stores set up vegetable stands outdoors.

In some cities, small stores sell vegetables. Sometimes many farmers gather to sell vegetables at an outdoor market.

Storing Vegetables

Vegetables can be stored in a number of ways. Many vegetables can be canned or **preserved** in jars. Some can be **pickled** or turned into juice.

Many fresh vegetables should be kept in the refrigerator. Celery, lettuce, and other vegetables stay fresh longer when cold. But potatoes and onions can be kept in a dark, dry cupboard.

Raw Vegetables

Many vegetables can be eaten raw. Lettuce, celery, onions, broccoli, and carrots are crunchy. They are often used in salads. Some people put lettuce and onions on sandwiches.

The tomato is the fruit part of a plant. But it usually is not very sweet. Many people consider it a vegetable. They eat it raw or cooked. Cucumbers and peppers are other fruits that are considered vegetables.

Cooking Vegetables

Vegetables can be cooked in many ways. They can be steamed or **sautéed**. They can be grilled, boiled, baked, or roasted.

Vegetables in a bamboo basket can be steamed over hot water.

Stir-fried vegetables stay slightly crunchy.

Vegetables can be combined and stir-fried so that the flavors blend together. They can also be cooked with **herbs** and spices to add to their flavor.

In many Asian countries, people eat corn before it is full grown. The tiny cobs and kernels are cooked in stir-fries and other dishes.

Baby corn cobs are quite small.

Russian beet soup is called borscht.

In Russia, beets are boiled into soup. In Germany, people eat sauerkraut with sausages. Sauerkraut is made from cabbage. French fries are popular in many countries.

The Vegetable Group

We should eat three to five servings of vegetables each day.

Glossary

bulb a round part of a plant that grows underground

carbohydrate an element found in certain foods that gives us energy when eaten; bananas, corn, potatoes, rice, and bread are high in carbohydrates.

chemical a substance used in science; chemical sprays or powders are sometimes put on plants to help them grow bigger or to keep pests away; organic farmers do not use chemicals.

greenhouse a building with walls and a roof made of glass so the sun can shine through; greenhouses can grow plants year round.

herb a plant that has leaves, stems, seeds, or roots that are used to flavor foods

mineral an element from earth that is found in certain foods; iron and calcium are minerals; we need small amounts of some minerals to stay healthy.

nutrient an element in food that living things need to stay healthy; proteins, minerals, and vitamins are nutrients.

pickle to soak in a salt or vinegar mixture; cucumbers soaked in a vinegar mixture are called pickles; pickled foods do not spoil as quickly.

preserve to fix food so that it will not spoil

protein an element found in certain foods that gives us energy when eaten; eggs, meat, cheese, and milk are high in protein.

sauté (saw-TAY) to cook food in a fat such as butter or oil

vitamin an element found in certain foods; Vitamin C is found in oranges and other foods; we need to eat foods with vitamins to stay healthy.

Index